Plants

Lesson 1

What Are Some Plant Types?..............2

Lesson 2

How Do Plants Grow?.....................10

Lesson 3

How Do Plants Reproduce?...............18

Lesson 4

How Do Plants of the Past
Compare with Those of Today?..............26

Orlando   Austin   New York   San Diego   Toronto   London

Visit *The Learning Site!*
www.harcourtschool.com

Lesson 1

VOCABULARY
classify
vascular
nonvascular

What Are Some Plant Types?

When you **classify**, you group things that are alike. The coin machine in the picture helps the boy classify coins by size.

Vascular means "having vessels." Vascular plants have tubes, or vessels. The tubes look like veins. They carry food and water around the plant. The tubes are easy to see in the leaves of plants.

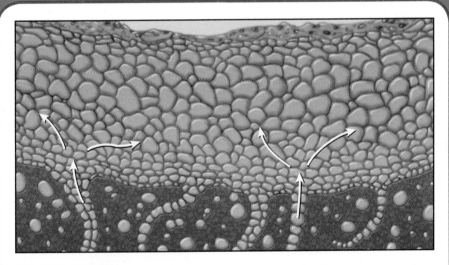

Nonvascular means "without vessels." Nonvascular plants do not have tubes, or vessels, to carry food and water. Each part of the plant takes in water. Nonvascular plants grow close to the ground.

READING FOCUS SKILL
COMPARE AND CONTRAST

When you **compare and contrast** you look for ways things are alike and different.

Look for different ways to **compare and contrast** the plants described in this lesson.

Trees and Grasses

When you **classify**, you group things that are alike. For example, scientists classify some plants as trees.

Some trees have leaves that change color and fall off in autumn. These are deciduous trees. Other trees have leaves shaped like needles that are green all year. These are evergreen trees. But all trees are tall, woody plants.

◀ The buckeye is a deciduous tree.

The pine tree is an evergreen. ▶

Wheat

Scientists classify some plants as grasses. Grasses have thin leaves. They are usually much shorter than trees. Wheat is a kind of grass. Turf is also a kind of grass. Turf is grass used for lawns.

 Compare trees and grasses.

Turf

Stems and Branches

You can classify plants by their stems. Most plants have stems. Some have soft, green stems. These stems bend easily. They have a thin covering of skin. Leaves and flowers may grow from them.

Some plants have woody stems. Woody stems are covered in bark. They do not bend. They can grow to be very thick.

◀ The lily has soft stems.

The rhododendron has woody stems. ▶

Oak tree

Pine tree

You can classify trees by the way their branches grow. Some trees have trunks that grow straight up. All the branches grow out of the trunk. Pine trees are like this.

Other trees have big branches that grow out of the trunk. Small branches grow out of the big ones. Oak trees are like this.

 Compare different kinds of stems and different kinds of branches.

Carrying Food and Water

All parts of a plant need food and water to live. Some plants have tubes, or vessels, that carry food and water to the different parts of the plant. You may be able to see the tubes in a leaf. They look like tiny lines.

Plants with tubes for carrying food and water are called vascular plants. **Vascular** means "having vessels."

Water and nutrients travel up the stem to the rest of the plant.

Other plants do not have tubes for carrying food and water. These plants grow close to the ground. They take in water from the soil, like sponges.

Plants without tubes are called nonvascular plants. **Nonvascular** means "without vessels."

 Tell how vascular and nonvascular plants are alike and different.

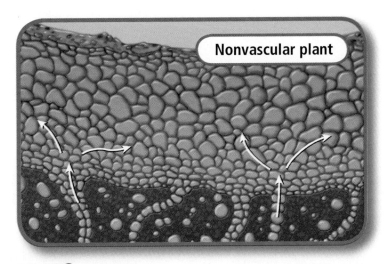

Nonvascular plant

Review

Finish these sentences that compare and contrast ideas.

1. Scientists classify pine trees and buckeyes as trees because they are tall and _____.

2. Two groups of stems include woody stems and _____ stems.

3. Vascular plants have _____ that carry food and water, but nonvascular plants do not.

4. Vascular and nonvascular plants are alike because they both need food and _____.

Lesson 2

How Do Plants Grow?

VOCABULARY
vascular tissue
xylem
phloem
photosynthesis

Vascular tissue holds up a plant. Vascular tissue also carries food and water to all parts of the plant. Roots, stems, and leaves have vascular tissue. This tree trunk has vascular tissue.

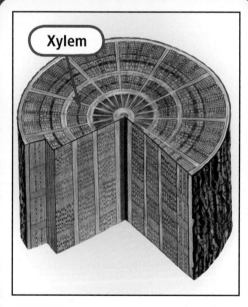

Xylem

Xylem is a kind of vascular tissue. Xylem carries water and nutrients from the roots to other parts of the plant.

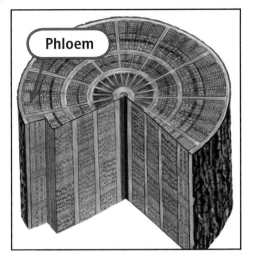
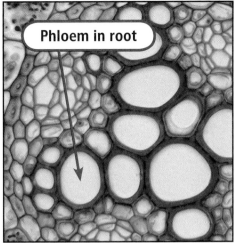

Phloem is a kind of vascular tissue. Phloem carries food from the leaves to the rest of the plant.

Photosynthesis is the process plants use to make food. Photosynthesis takes place in the cells of a plant's leaves.

READING FOCUS SKILL
MAIN IDEA AND DETAILS

The **main idea** is what the text is mostly about.
Details give added information about the **main idea**.
Look for **details** about how the different parts of a plant work to help the plant survive.

Nonvascular and Vascular Plants

Mosses are nonvascular plants. That means they do not have tubes for carrying food and water to different parts of the plant. They cannot grow tall.

Nonvascular plants do not have roots or stems like vascular plants. They do not even really have leaves. These plants take in water from their surroundings through small parts. The water carries food from cell to cell.

Trees are a kind of vascular plant. Vascular plants have vascular tissue. **Vascular tissue** holds up plants. It carries water and food. Roots, stems, and leaves have vascular tissue.

There are two kinds of vascular tissue. **Xylem** carries water and nutrients from the roots to the rest of the plant. **Phloem** carries food from leaves to the rest of the plant.

 Tell one detail about each kind of vascular tissue.

Each year new cells grow. You can tell the age of a tree by counting the rings of xylem.

Roots and Stems

Vascular plants have roots. Most roots have root hairs. Root hairs take in water and nutrients. Xylem cells carry the water and nutrients from the root hairs to the stem.

Roots helps hold a plant in place. Some plants have a taproot. A taproot is long. It grows deep into the soil. Some taproots also store food for the plant. Carrots are taproots that people eat.

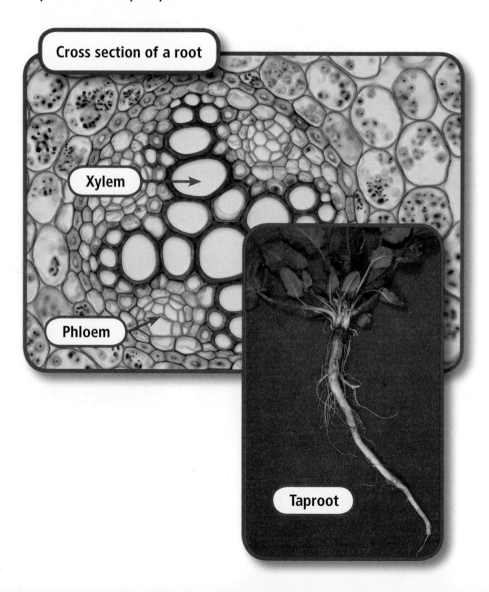

Cross section of a root

Xylem

Phloem

Taproot

Fibrous roots

Some plants have fibrous roots. Firbrous roots are thin. They branch out just below the surface of the ground. Plants with fibrous roots, like grass, help hold the soil in place.

Like roots, stems carry water and nutrients. Stems also hold a plant up. They hold the leaves out in the sunlight. Some plants, like cacti, store food and water in their stems.

 Tell details about how roots and stems help a plant.

Leaves

Leaves make food for a plant. The process used in making food is called **photosynthesis**. Light, carbon dioxide, and water are needed for photosynthesis. The food the plant makes is a kind of sugar.

Photosynthesis takes place inside the cells of a leaf. There, a green pigment takes in sunlight. Xylem cells carry water to the leaf. The needed carbon dioxide comes from the air. Phloem cells carry sugar to all parts of the plant.

Cross section of a leaf ▼

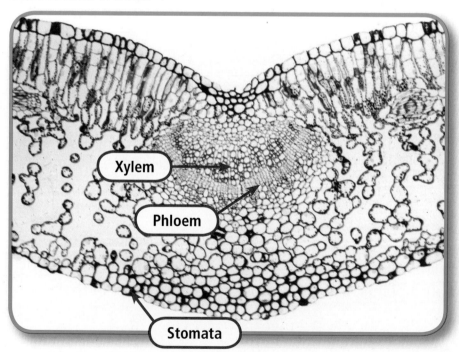

Leaves have an outer layer of cells called the epidermis. This layer of cells protects the leaf. It may have a waxy coating. The coating keeps moisture in the leaf.

On the underside of a leaf are tiny holes called stomata. Stomata open and close. They let carbon dioxide in. They let oxygen out. Some water may also pass through the stomata.

 Tell details about the process of photosynthesis.

Photosynthesis takes place in the leaves of a plant.

Review

Finish this main idea sentence.

1. Vascular and nonvascular plants make food by a process called _____.

Finish these details sentences.

2. The vascular tissue that carries water and nutrients is _____.

3. The vascular tissue that carries food to all parts of the plant is _____.

4. Taproots grow deep and _____ roots spread out near the surface.

Lesson

How Do Plants Reproduce?

VOCABULARY
spore
gymnosperm
angiosperm
germinate

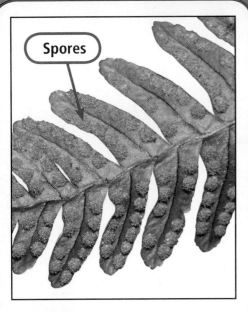

A **spore** is a cell that can grow into a new plant. Mosses are plants that reproduce by spores. Ferns also reproduce by spores.

A **gymnosperm** is a plant whose seeds are not protected by a fruit. A pine tree is an example of a gymnosperm.

An **angiosperm** is a flowering plant with seeds that are protected by a fruit. An apple tree is an example of an angiosperm.

To **germinate** means to begin to grow, or sprout. Most seeds need water and warmth to germinate. The seed in the photo is germinating.

READING FOCUS SKILL
COMPARE AND CONTRAST

To compare and contrast is to look for ways things are alike and different from one another.

Compare and contrast the different ways plants reproduce.

Reproducing by Spores

Some plants reproduce by spores. A **spore** is a cell that can grow into a new plant. Mosses reproduce by spores. Ferns also reproduce by spores. The life cycles of ferns and mosses are very much alike.

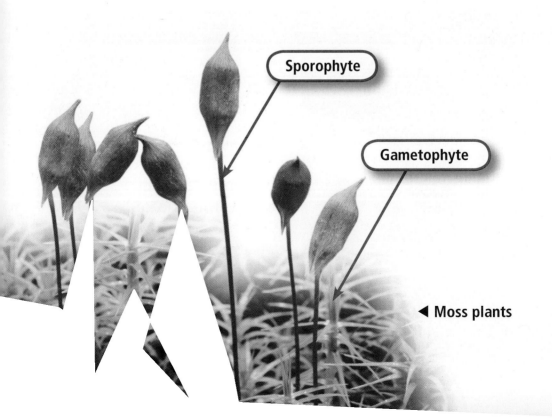

◄ Moss plants

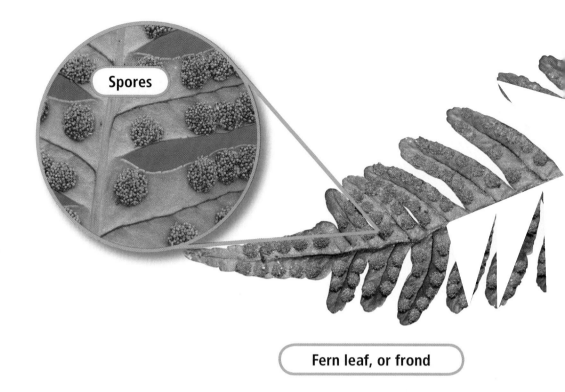

Spores

Fern leaf, or frond

First, the plant makes spores. The spores may be carried away by wind. Or, they may be let go in another way. The spores may grow to be gametophyte plants.

Male gametophyte plants make sperm. Female gametophtye plants make eggs. Egg and sperm join. Then a new plant grows. The new plant is a sporophyte. It makes spores.

 Tell a way fern and moss plants are alike.

Reproducing By Seeds

Many plants reproduce by seeds. Some plants, such as pine trees, produce "naked" seeds. Naked seeds are protected only by a seed coat. They do not grow inside a fruit as other seeds do. A plant that produces naked seeds is called a **gymnosperm**.

The seeds of a gymnosperm grow inside a cone. When the seeds are ready, the scales of the cone separate. Then, the seeds blow away. Some seeds may grow to become new trees.

▲ **Gymnosperm cone with seeds**

Many plants produce seeds that grow inside a fruit. These plants are called **angiosperms**. An apple tree is an angiosperm. It has flowers instead of cones.

The seeds of an angiosperm grow deep in the flower. Part of the flower becomes a seed coat. Another part becomes the fruit.

 Tell how the seeds of gymnosperms and angiosperms are different.

Seeds form here, in the ovule.

This part, the ovary, develops into a fruit.

Seed Germination

Most seeds need warmth and water to grow. Some kinds of seeds also need light. When everything is right, a seed may **germinate**, or begin to grow.

First the seed takes in water. The water makes the seed get bigger. Before long, the seed coat splits open.

A tiny root grows down into the soil. The root takes in water for the embryo, or young plant.

Germination

Next, a shoot pushes up. It grows toward the light. The leaves of the shoot cannot make food. The shoot uses food stored in the seed.

Then a stem grows. Roots spread out in the soil. Leaves form on the stem. Soon the leaves turn green.

The green pigment takes in sunlight. Then photosynthesis begins, and the plant starts to make its own food. At this point, the embryo is a plant seedling that is growing fast.

 Tell how the leaves of an embryo are different from the leaves of a seedling.

Seedling

Review

Finish these compare and contrast sentences.

1. Both ferns and mosses reproduce by means of _____.

2. The seeds of angiosperms are protected by a fruit, but gymnosperm seeds are _____.

3. Both angiosperm and gymnosperm seeds need warmth and _____ to germinate.

4. An embryo cannot make its own food, but a _____ can.

Lesson 4

How Do Plants of the Past Compare with Those of Today?

VOCABULARY
fossil
extinction

A **fossil** is what is left of a once-living thing. Many plant fossils are imprints, like the imprint of the fern plant in the photo.

Extinction is when all of a living species ~~dies~~. For example, the leaf in the photo is a fo~~ssil~~. It comes from a kind of plant that died out ~~lon~~g ago. The plant is extinct.

READING FOCUS SKILL
COMPARE AND CONTRAST

To **compare** is to look for ways things are alike. To **contrast** is to look for ways things are different.

Look for ways living plants are alike and different from plants of long ago.

Plants Without Fruit

Scientists study fossils of plants that lived long ago. **Fossils** are what is left of once-living things. The once-living things may be preserved by being petrified. Or, they may be preserved by leaving an imprint.

Many fossils come from plants that didn't have fruit. Ferns are an example. Fossil ferns are very much like ferns of today.

Fossil fern

Living fern

Fossils help scientists understand what the world was like long ago. For example, scientists found fossils in an arctic environment. Some were imprints of leaves. Others were petrified tree trunks. Today, no trees grow where the fossils were found. It is too cold and icy.

Compare the sago palm with this fossil of a frond. They are very similar.

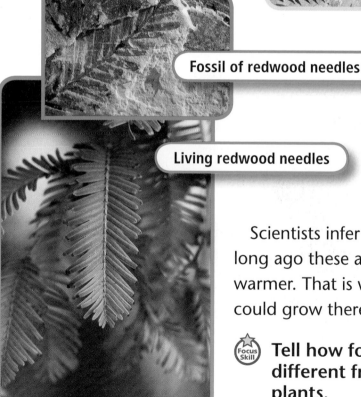

Fossil of redwood needles

Living redwood needles

Scientists infer that long ago these areas were warmer. That is why trees could grow there.

Tell how fossils are different from living plants.

Plants with Fruit

The fruit of a plant may be eaten by an animal. Or, it may rot away. That is why fossil plants that include fruit are not often found. When these fossils are found, they can teach scientists many things.

For example, scientists have found a fossil magnolia. The fossil includes a magnolia fruit. This shows that prehistoric magnolia had fruit just like magnolias of today.

Water lily fossils also show how environments have changed. Water lilies are easily damaged. An environment must have changed fast in order for a lily to become a fossil instead of being destroyed.

Fossils also help scientists learn about plants that have died out. One fossil plant had thin stems that reached above the water it grew in. It is the oldest known fruit-bearing plant.

Living waterlily

Fossil waterlily

 Compare lilies of today with fossil lilies.

Extinct Plants

Some prehistoric plants are not like any plants living today. Fossils are the only way scientists can learn about these plants. That is because the plants are extinct. **Extinction** is when all of a living species dies.

Scientists compare fossils of extinct plants with living plants. For example, the fossil on this page comes from an extinct tree. Scientists believe it is related to a modern tree.

 Tell how fossils help scientists learn about extinct plants.

Hymenaea protera

Finish these compare and contrast sentences.

1. A living plant may grow and change, but a _____ of a living plant will not.

2. Scientists have found fossils of trees in areas where it is too _____ for trees to grow now.

3. There are fewer fossils of plants with _____ than other kinds of plant fossils.

4. Scientists compare living plants with fossils of _____ plants.

31

GLOSSARY

angiosperm a flowering plant with seeds that are protected by a fruit, p. 23

classify group things that are alike, p. 4

extinction when all of a living species dies, p. 31

fossil what is left of a once-living thing, p. 28

germinate to begin to grow, or sprout, p. 24

gymnosperm a plant whose seeds are not protected by a fruit, p. 22

nonvascular "without vessels," Nonvascular plants do not have tubes, or vessels, to carry food and water. p. 9

phloem the vascular tissue that carries food from the leaves to the rest of the plant, p. 13

photosynthesis the process plants use to make food, p. 16

spore a cell that can grow into a new plant, p. 20

vascular "having vessels," Vascular plants have tubes, or vessels that carry food and water around the plant. p. 8

vascular tissue the part of a plant that holds up the plant and carries food and water to all parts of the plant, p. 13

xylem the vascular tissue that carries water and nutrients from the roots to other parts of the plant, p. 13